20 THINGS YOU DIDN'T KNOW ABOUT

FOSSILS

MARIE MORRISON

PowerKiDS press

Published in 2023 by The Rosen Publishing Group, Inc.
2544 Clinton Street, Buffalo, NY 14224

Portions of this work were originally authored by Janey Levy and published as *20 Fun Facts About Fossils*.
All new material in this edition was authored by Maria Morrison.

Editor: Greg Roza
Book Design: Tanya Dellaccio

Photo Credits: Cover Budimir Jevtic/Shutterstock.com; p. 5 Joaquin Corbalan P/Shutterstock.com; p. 6 AJancso/Shutterstock.com; p. 7 Sarunyu L/Shutterstock.com; p. 8 Renata Sedmakova/Shutterstock.com; p. 9 michal812/Shutterstock.com; p. 10 (fossil) servickuz/Shutterstock.com; p. 10 (seaweed) Zhane Luk/Shutterstock.com; p. 11 https://upload.wikimedia.org/wikipedia/commons/b/bc/Kampecaris_forfarensis%2C_Peach%2C_1882_01.jpg; p. 12 Victor1153/Shutterstock.com; p. 13 https://upload.wikimedia.org/wikipedia/commons/8/87/Megalosaurus%2C_World_Museum_Liverpool_%282%29.JPG; p. 14 diy13/Shutterstock.com; p. 15 Mark Brandon/shutterstock.com; p. 16 https://upload.wikimedia.org/wikipedia/commons/4/45/FMNH_Patagotitan.jpg; p. 17 (rhino) AKKHARAT JARUSILAWONG/Shutterstock.com; p. 17 (Spinosaurus) Warpaint/Shutterstock.com; p. 17 (blue whale) Maria Spb/Shutterstock.com; p. 17 (elephant) Art Konovalov/Shutterstock.com; p. 17 (Titanosaurus) Dotted Yeti/Shutterstock.com; p. 18 Dallas Golden/Shuttertock.com; p. 19 AKKHARAT JARUSILAWONG/Shutterstock.com; p. 20 W. Scott McGill/Shutterstock.com; p. 21 REUTERS/Alamy Stock Photo; p. 22 COULANGES/Shuttersstock.com; p. 23 high fliers/Shutterstock.com; p. 24 https://upload.wikimedia.org/wikipedia/commons/3/31/Tanis_site_photograph.jpg; p. 25 Ad_hominem/Shutterstock.com; p. 26 https://upload.wikimedia.org/wikipedia/commons/8/83/Fighting_dinosaurs_%281%29.jpg; p. 27 netsuthep/Shutterstock.com; p. 29 Kriengsak Wiriyakrieng/Shutterstock.com.

Cataloging-in-Publication Data
Names: Morrison, Marie.
Title: 20 Things You Didn't Know About Fossils / Marie Morrison.
Description: New York : PowerKids Press, 2023. | Series: Did You Know? Earth Science| Includes glossary and index.
Identifiers: ISBN 9781538389683 (pbk.) | ISBN 9781538389706 (library bound) | ISBN 9781538389713 (ebook)
Subjects: LCSH: Fossils–Juvenile literature. | Paleontology–Juvenile literature.
Classification: LCC QE714.5 M67 2023 | DDC 560-dc23

Manufactured in the United States of America

CPSIA Compliance Information: Batch #CWPK23. For Further Information contact Rosen Publishing at 1-800-237-9932.

CONTENTS

A PEEK INTO THE PAST

Dinosaurs and many other creatures lived long ago. So how do we know anything about them? Fossils! Fossils are the remains or traces of plants or animals that have been **preserved** in Earth's crust.

People can learn a lot from fossils. We can learn about where plants and animals lived thousands or millions of years ago. We can learn about how they lived and how they adapted, or changed to live better in their **environment**. We can even learn things about how they moved or had babies!

People have found fossils all over the world!

LEGENDARY!

DID YOU KNOW?

Dragon **legends** may have started with dinosaur fossils!

Chang Qu, a Chinese historian in the fourth century BCE, studied a fossil found in what's now the Sichuan Province. He decided it was a dragon! There are many dragons in Chinese legend.

Most Chinese dragons have long bodies like snakes. They're said to live in the water.

Long-ago gold miners in the Gobi Desert in Asia may have seen *Protoceratops* fossils and thought they came from griffins, legendary creatures that were part eagle, part lion.

FROM LONG AGO

DID YOU KNOW?

The earliest records of people studying fossils come from the eighth century BCE. That's nearly 3,000 years ago!

Scientists who study the past using fossils are called paleontologists. We know there were early paleontologists in China and ancient Greece.

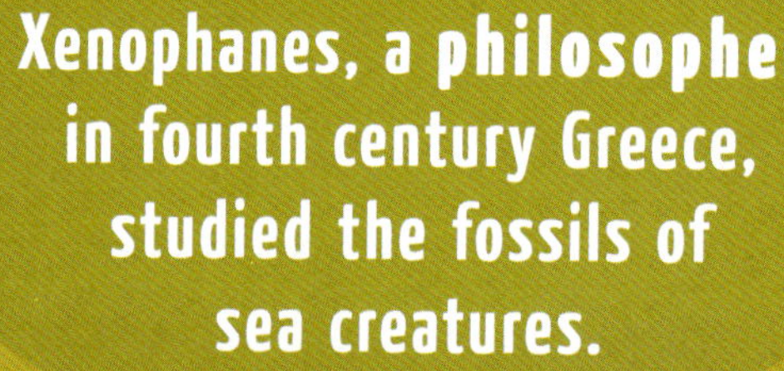

Xenophanes, a **philosopher** in fourth century Greece, studied the fossils of sea creatures.

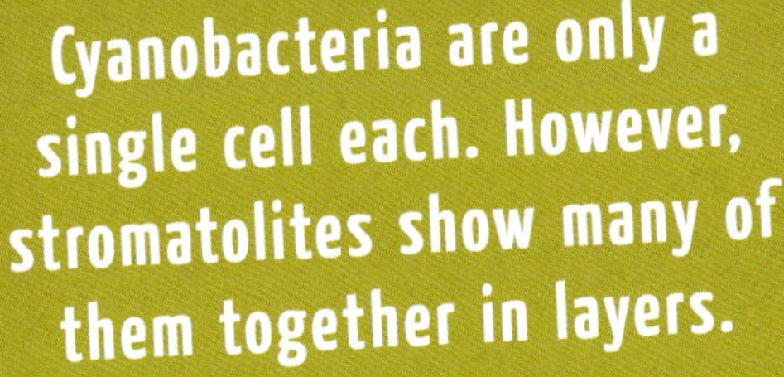

DID YOU KNOW?

The oldest known fossils are also some of the smallest.

People have found fossils that are about 3.5 billion years old in western Australia. These fossils are called stromatolites. They've made from tiny bacteria called cyanobacteria.

DID YOU KNOW?

The oldest known fossil of a green plant is about 1 billion years old.

In 2020, scientists discovered an ancient seaweed fossil in rocks in northern China. The small plants probably grew on the seafloor many years ago.

Seaweed can also be very large, but the ancient seaweed found in China was only about the size of a grain of rice.

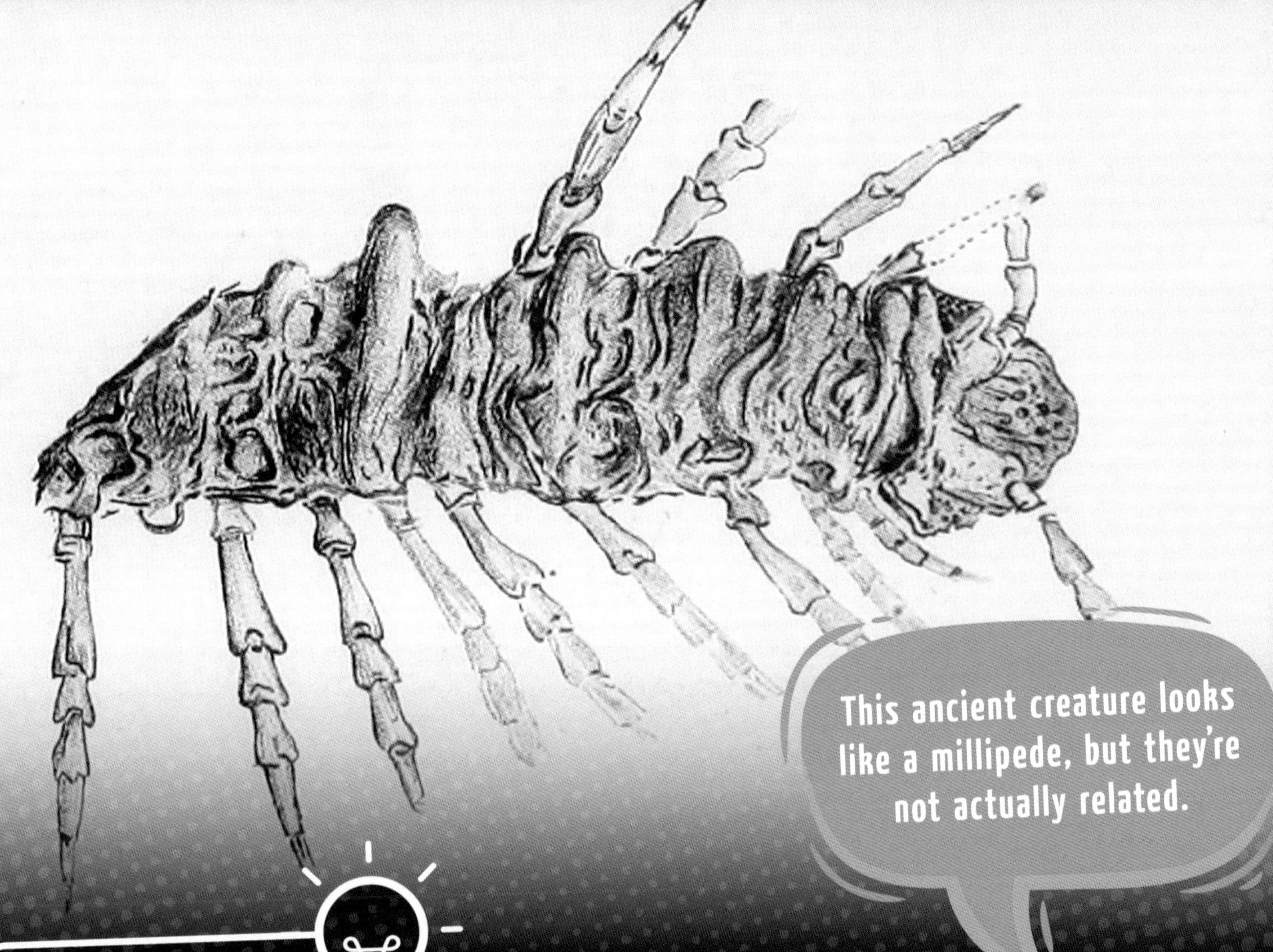

DID YOU KNOW?

The oldest land animal known from a fossil looks a lot like a modern **millipede**.

Researchers found a fossil of *Kampecaris obanensis*, a **segmented** creature, in 2020 in the Scottish Hebrides Islands. It's from about 425 million years ago.

AGE OF THE DINOSAURS

DID YOU KNOW?

Dinosaurs didn't get their name until 1842.

Richard Owen of England was studying some recently discovered fossils. He called the creatures "terrible lizards." "Deinos" is Greek for "inspiring fear" and "sauros" means "lizard."

One of the fossils Owen studied was that of an iguanodon. It was discovered in southern England.

About 20 years before Richard Owen named his fossils, William Buckland looked at some ancient bones and named the creature that'd owned them *Megalosaurus*. Buckland thought it was just an **extinct** giant lizard.

DID YOU KNOW?

Fossils show that most dinosaurs, unlike modern **reptiles**, were likely warm blooded.

In 2022, a study looked at fossil clues and concluded that many dinosaurs were more like modern birds and **mammals**. However, some, like stegosaurus, were cold blooded.

"Warm blooded" means that animals could keep their bodies warm even when it was cold around them. Cold blooded animals can't.

It wasn't until the 1990s that scientists figured out that many dinosaurs were feathered. Some scientists found a dinosaur fossil with feathers on its head, neck, back, and tail. Today we know older creatures probably had feathers too.

THEY'RE HUGE!

DID YOU KNOW?

Paleontologists keep finding bigger fossils!

It's hard to know what the biggest dinosaur ever was, because people keep finding ones that might be bigger! Also, most fossils aren't complete, so scientists must estimate, or take an educated guess, at their total size.

Titanosaurs were the biggest kind of dinosaurs. These herbivores, or plant eaters, could be more than 120 feet (36.6 m) long and weigh nearly 80 tons (72.8 mt)!

WHO'S THE BIGGEST?

While titanosaurs were huge, a creature living today is even larger! Who's the biggest of the big?

BLUE WHALE
(still living)

largest creature ever

About 150 tons (136.1 mt) and about 100 feet (30.5 m) long

TITANOSAURS
(extinct)

largest dinosaurs

About 80 tons (72.8 mt) and more than 120 feet (36.6 m) long

GIANT RHINOS
(extinct)

largest land mammals ever

More than 20 tons (18.1 mt) and more than 20 feet (6.1 m) high at the shoulder

AFRICAN BUSH ELEPHANT
(still living)

largest land mammal

Up to 13,000 pounds (5,896 kg) and 24 feet (7.3 m) long

SPINOSAURUS
(extinct)

largest land predator ever

Up to 22 tons (20 mt) and about 50 feet (15.2 m) long

DID YOU KNOW?

Paleontologists have found fossils of a huge "dragon of death"!

Thanatosdrakon is a kind of pterosaur, or extinct flying reptile. That name means "dragon of death"! These creatures had a wingspan of up to 30 feet (9.1 m) wide.

Pterosaurs could also be very small—as little as the birds you see outside!

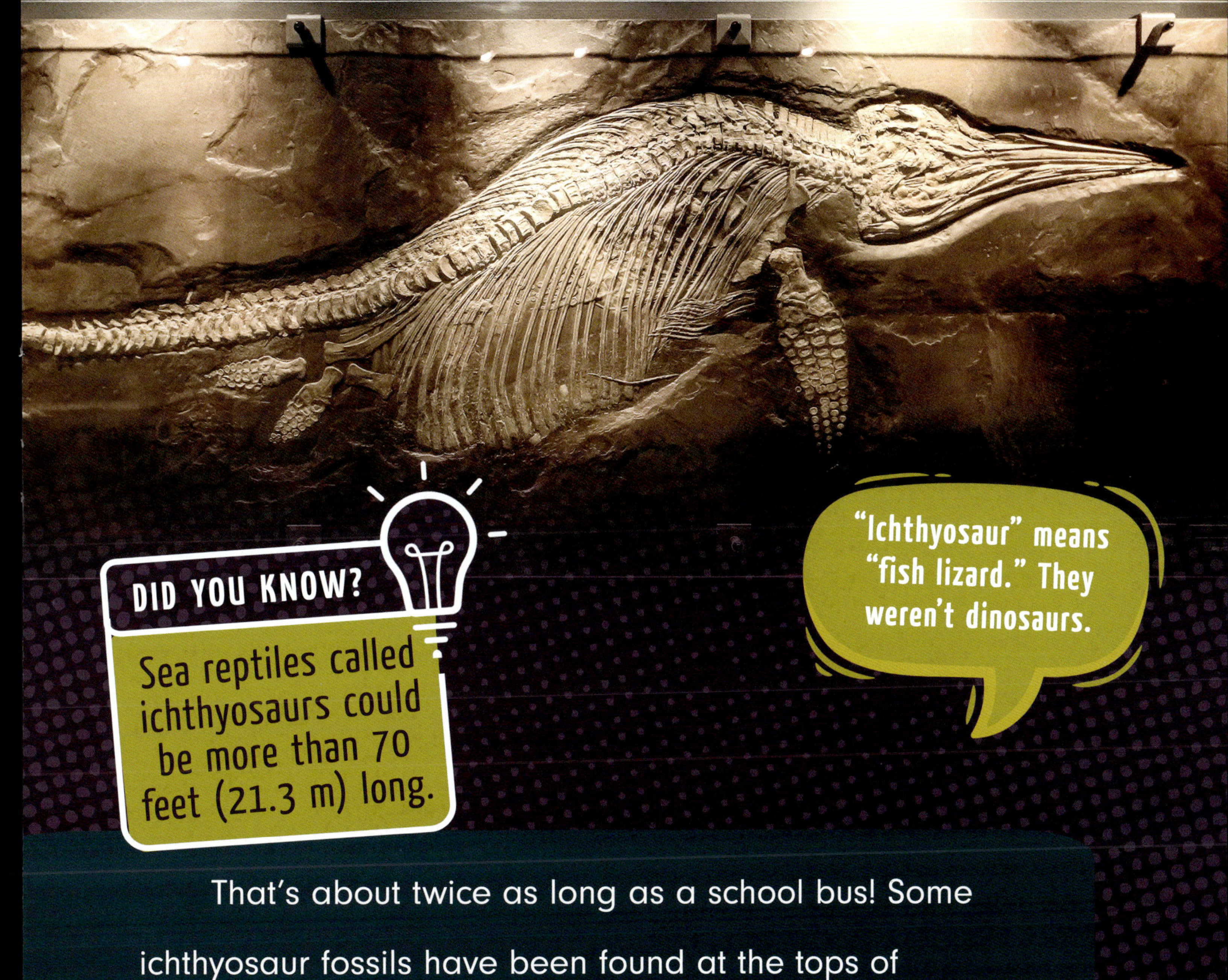

DID YOU KNOW?

Sea reptiles called ichthyosaurs could be more than 70 feet (21.3 m) long.

That's about twice as long as a school bus! Some ichthyosaur fossils have been found at the tops of mountains. Those mountaintops used to be seafloors.

RAPTOR REALITY

DID YOU KNOW?

Fossils show us that velociraptors were fairly small and had feathers.

These dinosaurs may look big and scary in movies, but they were really only about as big as a wolf—about 100 pounds (45.4 kg). They might have had bright colors.

Velociraptors had a big, curved claw on each foot.

In 2022, paleontologists found fossils of the biggest raptor discovered so far. It was about 33 feet (10 m) long and had claws nearly 16 inches (40.6 cm) long.

PLANTS LEAVE FOSSILS TOO

DID YOU KNOW?

Fossil plants can help us learn about **climate change.**

Scientists can study the leaves of fossil plants (and other features) to learn what sort of climate the plants grew in. This can also show how plants may deal with climate change today.

Palm trees almost always grow in tropical, or warm and wet, climates, so palm fossils likely mean that plant grew in a similar climate.

DID YOU KNOW?

Scientists have found fossilized fruit that's about 52 million years old.

You wouldn't want to have this fruit for a snack! Paleontologists found the fossils in South America. The ancient berry belonged to the family that includes tomatoes and peppers.

COOL FOSSIL FINDS

DID YOU KNOW?

Newer fossil finds may show dinosaurs that were killed in the event that caused the creatures' extinction.

Many scientists believe that dinosaurs went extinct when an **asteroid** crashed into Earth about 66 million years ago. These fossils may show direct results of that crash.

The fossils were found at the Tanis fossil site in North Dakota. One fossil shows a perfectly preserved dinosaur leg, complete with skin.

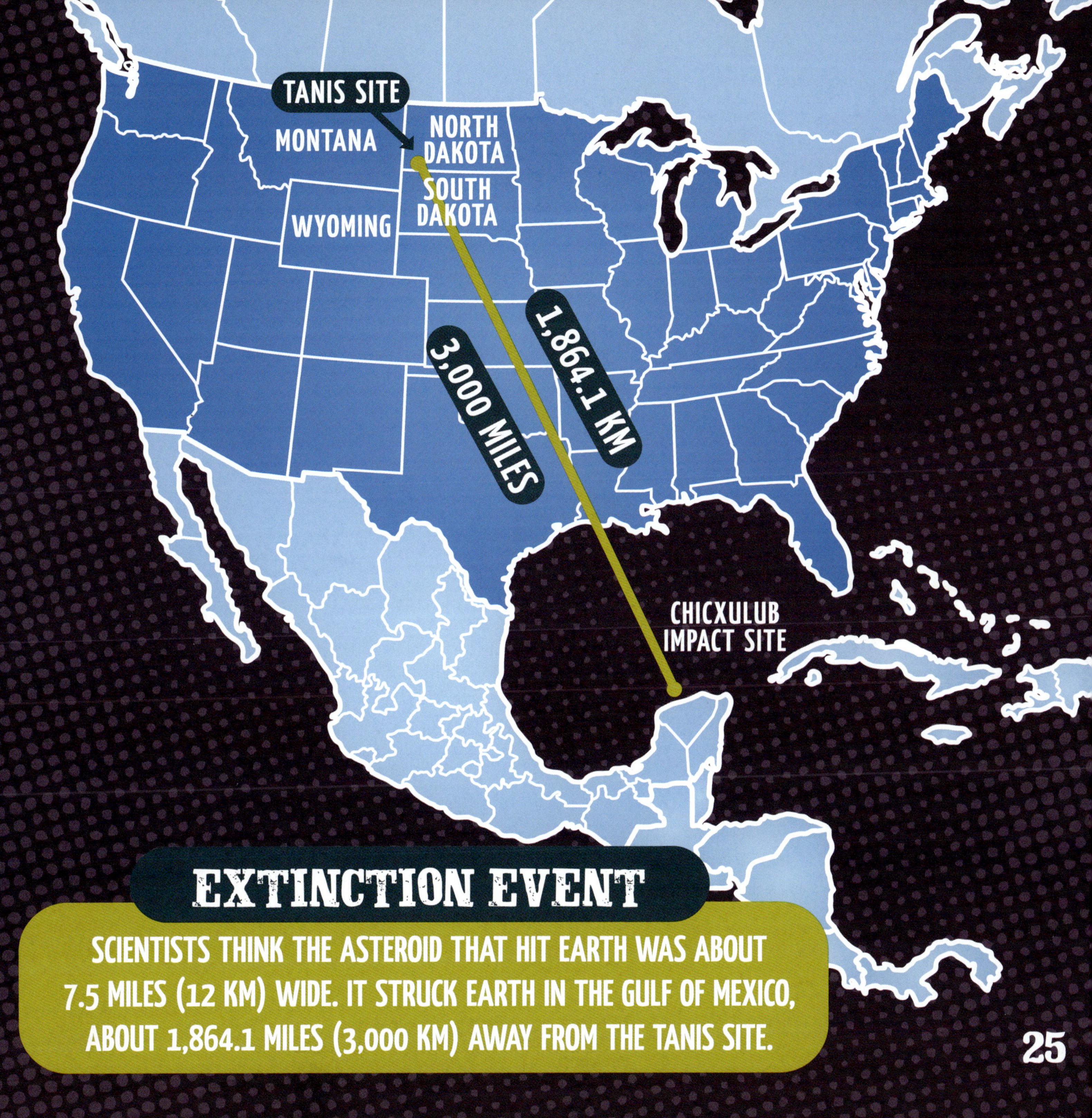

EXTINCTION EVENT

SCIENTISTS THINK THE ASTEROID THAT HIT EARTH WAS ABOUT 7.5 MILES (12 KM) WIDE. IT STRUCK EARTH IN THE GULF OF MEXICO, ABOUT 1,864.1 MILES (3,000 KM) AWAY FROM THE TANIS SITE.

One incredible fossil shows two dinosaurs fighting each other.

This fossil, found in Mongolia, shows a velociraptor and a protoceratops fighting in the Gobi Desert. In the middle of battle, a hill of sand fell on them!

This fossil, found in 1971, gave scientists a lot of information about both kinds of dinosaurs.

In 2021, a four-year-old girl in Wales found a dinosaur footprint on a beach. The footprint may be about 220 million years old! It's now in a museum.

FOSSILS OF THE FUTURE

Scientists and others are finding new fossils every day! Every fossil gives us a chance to learn more about the creatures and plants that have lived on Earth before us—and even the ones that are still here.

You can find fossils too! Look around sandy or muddy areas or places where there is old rock. Never go fossil hunting alone, and take notes on what you find. Maybe you'll find something no one's ever seen before.

Looking for fossils can be like traveling back in time.

GLOSSARY

asteroid: A small, rocky body in space.

climate change: Change in Earth's weather caused by human activity.

environment: The conditions that surround a living thing and affect the way it lives.

extinct: No longer existing.

genus: The scientific name for a group of plants or animals that share most features.

legend: A story coming down from the past that is popularly accepted but can't be checked.

mammal: Any warm-blooded animal whose babies drink milk and whose body is covered with hair or fur.

millipede: A bug having a long, segmented body and many pairs of legs.

philosopher: A person who studies philosophy, or the study of the basic ideas about knowledge, right and wrong, reasoning, and the value of things.

preserved: Kept from being destroyed.

reptile: A cold-blooded animal with thin, dry pieces of skin called scales.

segmented: Having segments, parts into which something can be divided.

FOR MORE INFORMATION

BOOKS

Hall, Ashley. *Fossils for Kids: A Junior Scientist's Guide to Dinosaur Bones, Ancient Animals, and Prehistoric Life on Earth.* Emeryville, CA: Rockridge Press, 2020.

Lomax, Dean. *My Book of Fossils.* New York, NY: DK Publishing, 2022.

Lynch, Dan. *Fossils for Kids: Finding, Identifying, and Collecting.* Cambridge, MN: Adventure Publications, 2020.

WEBSITES

Earth Science for Kids: Fossils
www.ducksters.com/science/earth_science/fossils.php
Ducksters has information about fossils and other related topics.

Fossils
www.dkfindout.com/us/dinosaurs-and-prehistoric-life/fossils/
This DK Find Out website offers many facts about fossils.

Paleontology: The Big Dig
www.amnh.org/explore/ology/paleontology
The American Museum of Natural History has a wealth of information on studying paleontology.

Publisher's note to educators and parents: Our editors have carefully reviewed these websites to ensure that they are suitable for students. Many websites change frequently, however, and we cannot guarantee that a site's future contents will continue to meet our high standards of quality and educational value. Be advised that students should be closely supervised whenever they access the internet.

INDEX